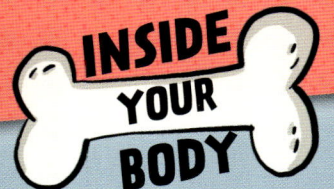

INSIDE
YOUR
BODY

GROWING

Angela Royston
Emiliano Migliardo

WAYLAND

First published in Great Britain in 2024 by Hodder & Stoughton.

Copyright © Hodder and Stoughton, 2024

The text in this book was previously published in the series 'Your Body Inside and Out'.

Series Editor: Grace Kelly
Designer: Lisa Peacock
Illustrator: Emiliano Migliardo

HB ISBN: 978 1 5263 2519 8
PB ISBN: 978 1 5263 2518 1

Wayland, an imprint of
Hachette Children's Group
Part of Hodder and Stoughton
Carmelite House
50 Victoria Embankment
London EC4Y 0DZ
An Hachette UK Company
www.hachette.co.uk
www.hachettechildrens.co.uk

Printed in China

MIX
Paper | Supporting
responsible forestry
FSC® C104740
FSC
www.fsc.org

CONTENTS

NOTE:
If you have any worries about changes to your body as you grow, please speak to a trusted adult such as a parent, carer or teacher.

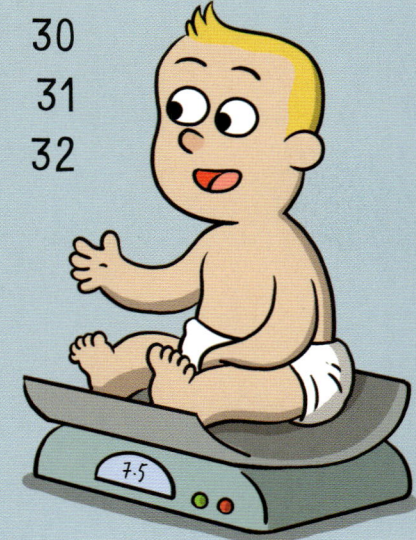

THE CYCLE OF LIFE

Humans grow and change a great deal during their lives. Each person starts life inside their mother. After nine months or so, a baby is born. The baby grows into a child, then a teenager, then an adult. Adults can produce children of their own. Eventually everyone dies. These changes are known as the human life cycle.

TRY THIS! Do you ever measure your height? It's interesting to see how you grow. Sometimes you hardly seem to grow at all. Then suddenly you shoot up quite quickly. Measure your height every month for a year (or more!). Track your changes on a chart.

YOU'VE HAD A MAJOR GROWTH SPURT!

Measuring your height is a good way to track how much you grow from year to year.

YOUR HEIGHT

Everyone grows up differently. Some people become tall and others don't. It's hard to say how tall you might grow. However, if there are a lot of tall people in your family, you'll probably be tall as well.

HIGH-FIVE!

LONGEST-LIVED PEOPLE

Studies show that Japanese people are one of sthe longest-lived people. They live for 85 years on average. They also have fewer illnesses than other people. One reason for this is probably the Japanese diet, which is very healthy.

PRODUCING BABIES

All living things reproduce. Birds, lizards, crocodiles and frogs lay eggs. The eggs hatch into baby birds or animals. Plants produce seeds that grow into new plants. Mammals, such as horses, cats and mice, give birth to live young. Humans are mammals, too.

HELLO WORLD!

Once baby lizards hatch, they are like miniature adults and don't need much help from their parents. They can walk, run and find food for themselves.

HUMAN REPRODUCTION

Every human has two parents, a father and a mother. We are related to our parents through our genes. We get half our genes from our mother and half from our father.

BILLIONS OF PEOPLE

Humans have been very successful at reproducing. The number of people in the world has grown and grown. In 1800, there were about one billion people in the world. Today there are over eight billion people!

SPECKS OF LIFE

Each human being begins life as a tiny egg inside a woman. A human egg is too small to see. It's about the width of a single hair.

FERTILISATION

Before it can grow into a baby, an egg must be fertilised. This means the egg joins up with a sperm cell from the father. The fertilised egg grows inside the mother's womb.

You can sometimes tell that people are related, because they look similar.

INSIDE THE WOMB

A baby grows inside its mother's womb. Her tummy stretches to make space for it. It takes about 40 weeks (9 months) for a baby to grow inside its mother.

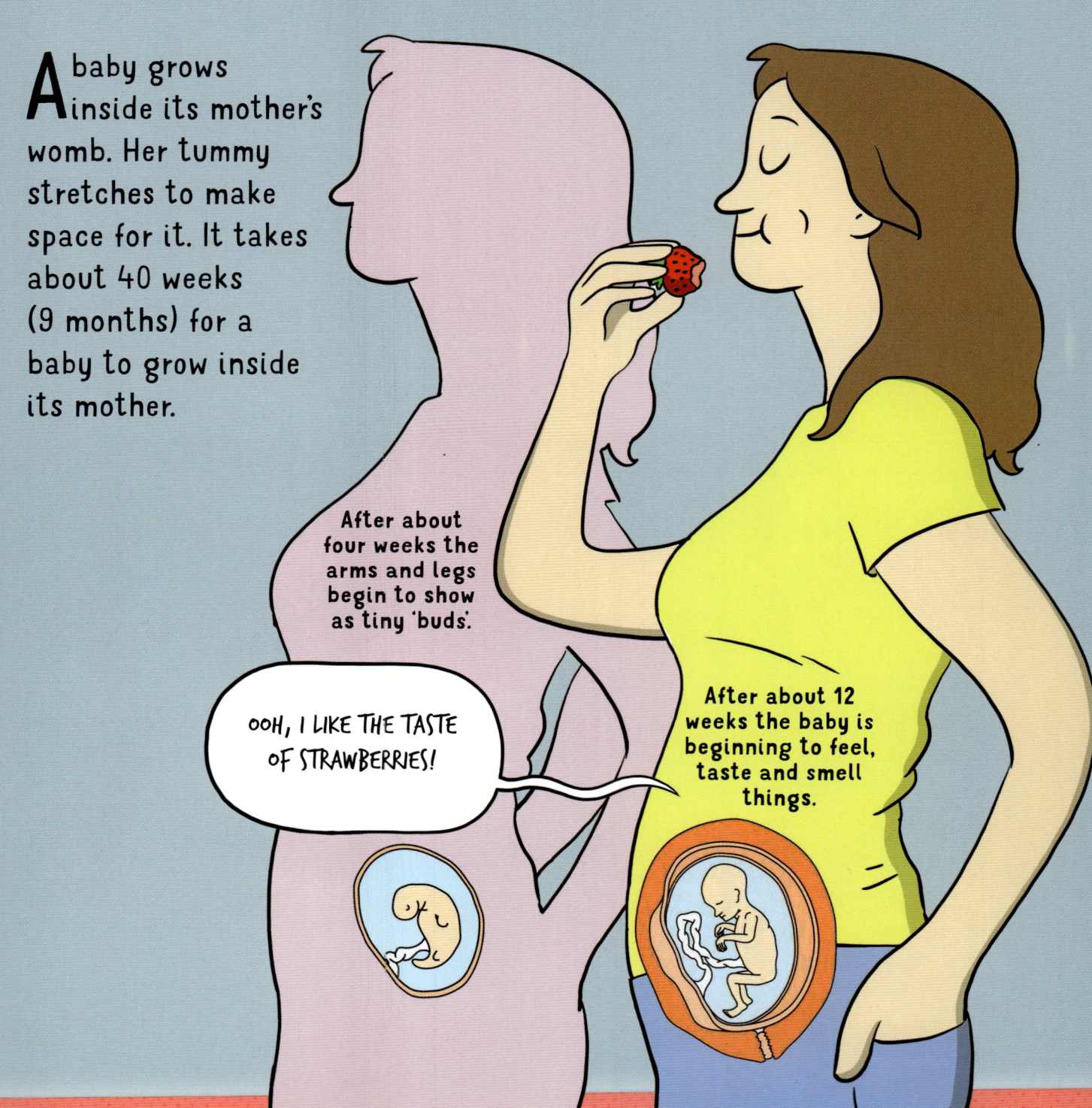

After about four weeks the arms and legs begin to show as tiny 'buds'.

OOH, I LIKE THE TASTE OF STRAWBERRIES!

After about 12 weeks the baby is beginning to feel, taste and smell things.

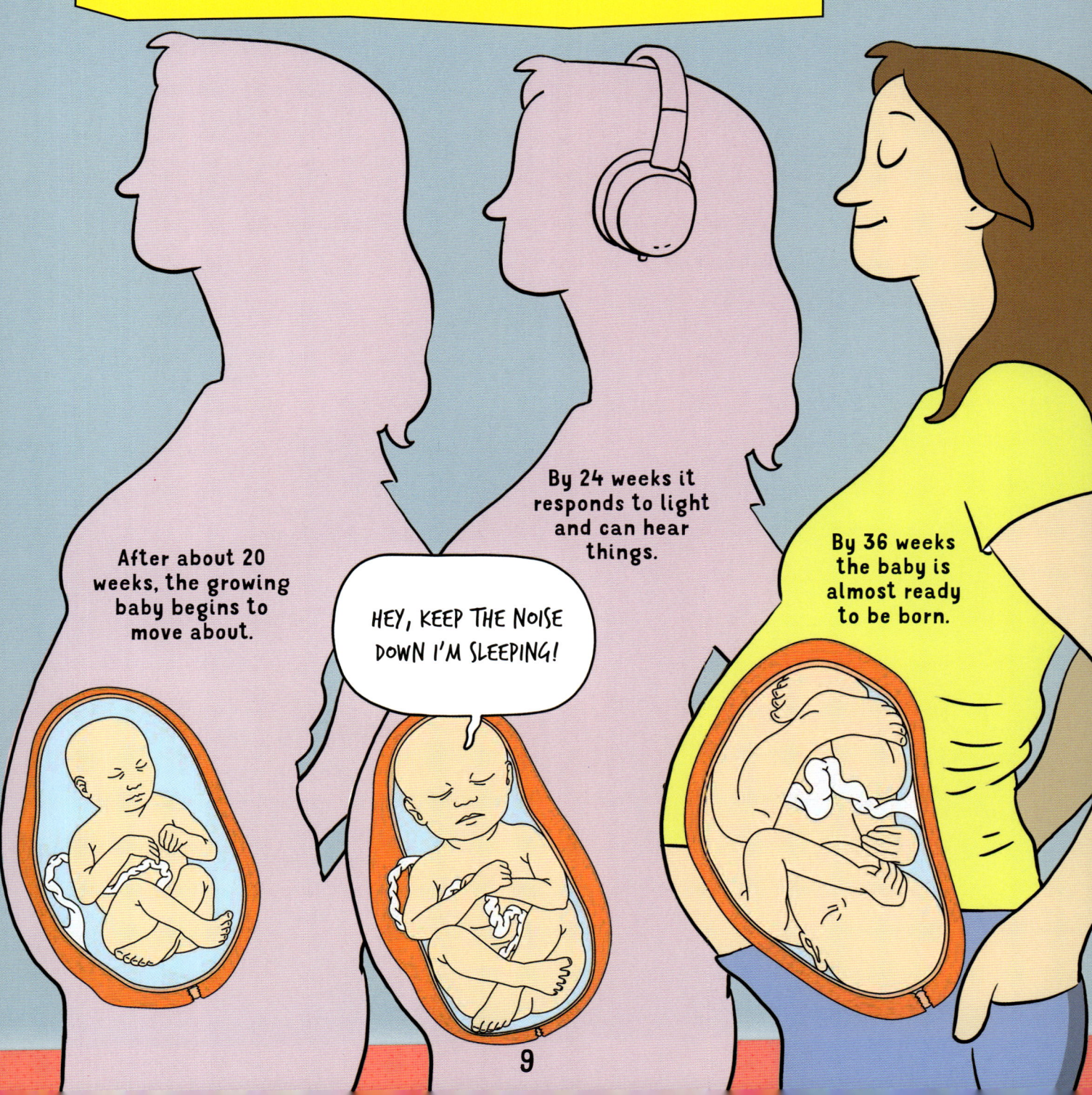

FLEXIBLE SKULL
A human skull is made of several different bones joined tightly together. In a new baby, the skull bones are only loosely joined. As the baby is being born, the bones can move slightly. This makes the birth easier.

After about 20 weeks, the growing baby begins to move about.

By 24 weeks it responds to light and can hear things.

HEY, KEEP THE NOISE DOWN I'M SLEEPING!

By 36 weeks the baby is almost ready to be born.

EVERYONE IS DIFFERENT

No two babies are born the same, and everyone grows and develops in different ways.

BABY WEIGHT

Babies vary a great deal in weight when they are born. A very big baby can weigh three times as much as a small one. A small newborn baby can weigh as little as 1.5 kilograms. This is the usual weight of a bag of flour. A really big baby can weigh over 4.5 kilograms. This is the same as three bags of flour. Both small and big babies usually grow up normally.

THAT'S FIVE BAGS OF FLOUR – GROWING NICELY!

FLOUR
FLOUR
FLOUR
FLOUR
FLOUR

7.5

WE HAVE DIFFERENT EYES!

TWINS, TRIPLETS AND MORE

Most mothers have one baby at a time. However, sometimes two, three or more babies develop inside the mother. Sometimes these babies are 'identical', sometimes not. However, even identical children have differences between them.

HAIR, EYES AND SKIN

Babies can have blonde hair, dark hair or red hair. Their skin can be light or dark. Their eyes could be any colour from light grey to black. We inherit these different characteristics from our parents, through our genes.

DEVELOPMENT

There are many other differences in the way that babies develop. One baby may walk at an early age. Another may take longer to walk but be good at learning words. These differences in how we develop continue throughout our lives.

YOU'RE TALL FOR AN EIGHT YEAR OLD!

A NEWBORN BABY

A newborn baby cannot move around. It cannot walk, crawl or even hold its head up. However it can move its arms and legs, and wriggle its body.

I CAN'T RUN YET, BUT I CAN WRIGGLE!

GROWING BONES

We all have a bony skeleton, but a newborn baby's skeleton is not completely made of bone. Large parts of the skeleton are made of a tough, springy material called cartilage. As a baby grows, the springy cartilage is replaced by stronger, harder bone.

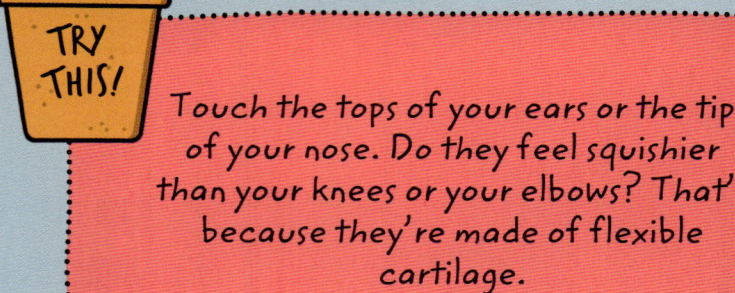

TRY THIS!

Touch the tops of your ears or the tip of your nose. Do they feel squishier than your knees or your elbows? That's because they're made of flexible cartilage.

12

EATING AND SLEEPING

Newborn babies spend most of their time eating and sleeping. They sleep for over 16 hours a day, in short naps. When newborn babies are hungry they wake up and feed. They cannot eat solid food. They drink milk, either from their mother's breast or from a bottle.

REST YOUR SLEEPY HEAD.

When you hold a newborn baby, you have to support its head. The baby's muscles are not strong enough to hold up its own head.

ARE YOU HUNGRY? TIRED?

COLD!

SIGHT AND SOUND

Newborn babies cannot see very well, but they have a good sense of smell. A baby recognises its mother mostly by smell. Babies cannot talk, but they can make a lot of noise! A baby cries to get the attention of its parents.

FROM SIX MONTHS TO ONE-YEAR-OLD

A baby grows very quickly in its first year, getting stronger and developing in many ways.

BY SIX MONTHS

A baby's eyesight soon improves, and it begins to look around. At six months most babies are sitting up with some support. They still drink mostly milk, but they may be eating some solid food, too.

I SEE IT, I WANT IT!

By around six months old a baby can usually hold its head up with some help. By nine months it can sit and reach for things that look interesting.

AT ONE YEAR

Around a baby's first birthday it will probably be crawling, standing or even walking. One-year-olds can hold a spoon, and are starting to feed themselves. They may also be able to say a few words. A one-year-old baby may weigh three times more than when it was born.

TODDLERS

Babies do not grow so much in their second year. However, their bodies change a lot. As a baby learns to walk, its muscles develop. Its legs and back get stronger, to help it stand upright. By the age of two, babies become toddlers.

DRAWING AND KICKING

As babies grow into toddlers, they learn to control their bodies better. By 18 months most toddlers are beginning to draw with crayons and by around two years old they can do things like build a tower of bricks and kick a ball.

This toddler can hold a crayon in his hand.

BA-NA-NA

TALKING

This is the time when children learn to talk more too. They start by listening and copying. Babies say single words first, but they soon start putting words together to make two-word sentences, such as 'me drink'.

This toddler is watching his mother's mouth as part of the process of learning to talk.

This illustration shows the parts of the brain that are active when we speak. The most active areas (shown darker) are on the left side of the brain.

TALKING AND THE BRAIN

From very early in life, we recognise talking as different from other sounds. The brain has two sides. Both sides process sounds but when we talk or listen to speech, we only use parts of the left side of the brain.

FROM TODDLER TO CHILD

At two, children mostly play by themselves. They do not understand how to play with other children. By four they are playing with other children and beginning to make friends.

TODAY I'M A DINOSAUR ...

By age four children start to become more independent.

BECOMING INDEPENDENT
By the age of four, children can do many things for themselves. They can climb the stairs, run, throw a ball and ride a tricycle. They learn to dress themselves and wash their hands. Instead of using two-word phrases, they begin to speak in sentences.

BRAINY CHILDREN

A child's brain learns and remembers much quicker than an adult's can. A child's brain and body work together to learn balance and co-ordination.

ONE, TWO, THREE ...

By the age of four most children are learning to count. However, they do not really understand numbers beyond two or three.

GROWING PAINS

At night, some children get pains in their leg muscles and joints. These are often called 'growing pains'. Doctors think these aches and pains may just be caused by being active during the day. Stretching or putting a heated pad on the painful area can help.

YOUNG CHILDREN

Between the ages of four and eight, children's bodies take on a more adult shape. Many new skills are also being learned.

CHANGING SHAPE

Babies have a large head and a short body. As we grow and begin to walk, the body, legs, and arms get longer, so the head seems smaller.

Age six months

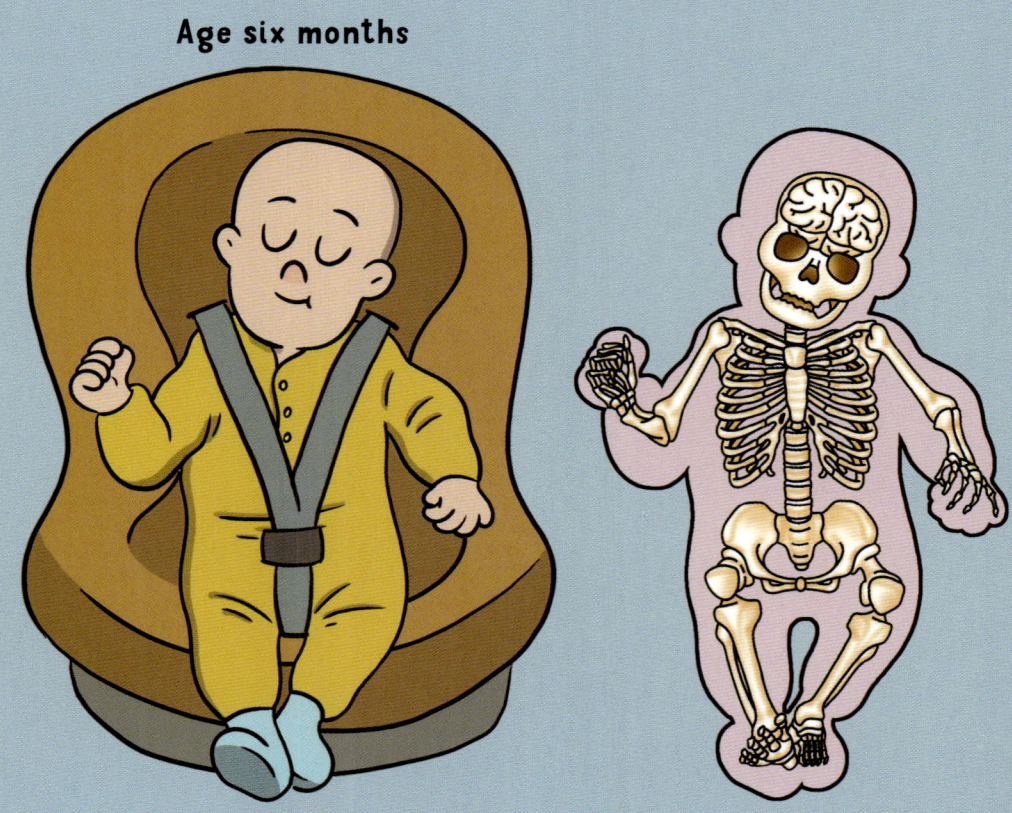

The body continues to grow more than the head as children get older. By the age of six the brain, and the head, have reached almost full size.

IN THE BRAIN

By six or seven most children can read and write. From age four to six, the part of the brain to do with language grows quickly. This is the best time to learn new languages. After the age of 12 we find learning a new language much harder.

HOW DO I SAY THANK YOU IN SPANISH?

Most children like to have stories read to them from an early age. By the age of six or seven, they can read by themselves, too.

Age six years

FROM CHILD TO ADULT

From about nine years old, young people begin to change from children to adults. This change is called puberty.

One change that happens in boys is that their 'Adam's apple' (the lump in their throat) gets bigger. This gives them a deeper voice.

I LIKE MY NEW VOICE!

GROWTH SPURT

The first sign of puberty is a growth spurt. The hands and feet grow first, then the arms and legs, and finally the body. Girls have this growth spurt before boys. Many girls reach their full height by around 13. Boys carry on growing into their late teens, and generally end up being taller than girls.

MALE AND FEMALE CHANGES

Other changes happen in puberty, too. Both boys and girls grow hair in new places. Girls' hips get wider, their breasts grow and they start their period. Boys' voices get deeper, hair grows on their face and they become taller and more muscular.

WHAT STARTS PUBERTY?

Hormones are special chemicals that start the change from child to adult. A hormone is produced in one part of the body and released into the blood. It then travels round in the blood and affects other parts of the body. The hormone that starts puberty is made by the pituitary gland in the brain.

WE'RE YOUR HORMONES, HELPING YOU GROW FROM CHILD TO ADULT!

STARTING A FAMILY

Once we go through puberty, we are able to have babies ourselves if we want to. Most people don't start a family until they are older.

The hormones produced during puberty can affect young people's emotions as well as their bodies. Teenagers sometimes feel upset or angry for no obvious reason.

SO MANY DIFFERENT FEELINGS ...

TRY THIS!

If you've had a growth spurt recently you might notice that some of your clothes are too small. Do you know someone you could give them to, or donate them to charity?

FULLY GROWN

M en and women grow to their full height in their late teens, but the body never really stops changing. Everyone matures differently so there is no set age when we are fully grown.

MUSCLE CHANGES
Physically, we're at our peak when we're teenagers and young adults. In late puberty, bones get stronger. There is also a 'strength spurt', especially in young men, when the muscles grow quickly.

Most of the world's top sportspeople are in their teens or twenties.

BULGING BRAINS

From the age of 12 to about 18, the front part of the brain enlarges. This is the thinking part of the brain. It is an important time for learning and organising new information. The knowledge and skills young people learn at this time become 'hard-wired'. They become a permanent part of the brain.

I'M LEARNING HOW TO COMPLETE THIS VIDEO GAME!

The changes in their brains make young adults good at learning and at organising new information. They are at a good stage in their life to face new challenges.

Skull

Front of brain (the thinking part)

GETTING OLDER

As adults get older, their bodies gradually work less well. Living a healthy lifestyle (see pages 28-29) can reduce the impact of ageing, but eventually everyone dies.

SIGNS OF AGEING

As we grow older our skin becomes less elastic, which causes wrinkles. Men often begin to lose their hair. Both men and women start to get grey hairs.

After middle age, people's eyes don't work as well. People who did not wear glasses when they were young start to need them for reading.

MUSCLES AND BONES

As people move into old age, their muscles work less well, and their bones become less strong. Bones reach their peak size and strength when people are in their mid-20s. After this, they slowly get thinner and less strong.

WHO ARE YOU CALLING WEAK?!

One effect of weaker bones is that, as people get older, their spines become more curved.

LIFE SPAN

How long we live depends on many factors – including how we look after ourselves, and the genes we inherit from our parents. People are living longer on average these days because of better healthcare and access to healthy food.

GOOD TIMES

Getting old is not all bad! Older people have more life experience. This helps them to solve problems and make difficult decisions. When their children leave home, older people often have more free time. They can enjoy themselves and take up new interests.

Taking up an activity such as walking is great for keeping bones and joints working well as people get older.

KEEPING HEALTHY

The best way to be healthy when you are older is to be active and eat a well-balanced diet throughout your life. This will help you stay healthy on the inside and out!

YOUR FOOD

Eat at least five portions of fresh fruit and vegetables a day. This will give you important vitamins and minerals, and plenty of fibre. Try not to eat too many sweet or fatty foods.

AIM TO EAT FIVE OF US A DAY!

If you eat healthy food all your life, you are more likely to stay healthy into old age.

KEEP ACTIVE

Keeping active is also important. Cycling, hiking, swimming and dancing are all good ways to keep your bones and muscles strong.

Dancing is good exercise for people of all ages, and it's fun!

USE YOUR BRAIN

It is also important to keep the brain active. Learning new things and solving problems will help you to keep your brain ticking over. If you look after your body and your mind, you have a better chance of living a long, healthy life!

TRY THIS!

Keep a record of your growing journey. Ask your family for photographs of you when you were a baby, a toddler and more recently. Keep them in an album or a scrapbook and track how much you've changed.

GLOSSARY

cartilage Strong, springy material that makes up some parts of the skeleton. As we grow, some of the cartilage in the skeleton is replaced by bone.

diet Everything that we eat.

fertilised When a woman's egg combines with a man's sperm, it starts to grow and develop into a baby.

genes Substances inside our bodies that act as 'instructions' for how we develop and grow. Genes are passed on from parents to their children.

growth spurt A time during puberty when young people grow very quickly.

hormones Substances that are produced in one part of the body and released into the blood. They travel through the blood and might help us to grow or affect our emotions.

human life-cycle The changes in a person's life as they are born, grow into adults and then have babies of their own.

identical Identical twins occur when the fertilised egg splits in half, so that two babies develop inside the mother. They will both be the same sex and look very similar.

life span the length of time that a person or animal lives.

minerals Simple chemicals such as iron and calcium, which we need in our diet.

puberty A period from about nine years old, when young people begin to change from children into adults.

period Also called menstruation, this happens to females who have gone through puberty. For a few days each month an egg is released and blood comes out of the vagina.

skeleton The framework of bones that supports our bodies.

sperm Tiny specks of life even smaller than a human egg. They are present in a milky liquid called semen that men produce. Sperm is needed to fertilise a human egg.

tummy Many people call the part of the body below the chest the tummy. The tummy includes the stomach, which digests food, and gets a little squashed as the womb and baby grow in size.

vitamins Substances that we need in our diet but only in small amounts.

womb The place inside a female body where a baby can grow and develop.

FURTHER INFORMATION

BOOKS

The Bright and Bold Human Body: The Reproductive System
by Sonya Newland (Wayland, 2020)

Me and My World: My Growing Body
by C.J. Polin (Franklin Watts, 2021)

What is DNA?
by Professor Julian Barwell and Dr. Neeta Lakhani (Wayland, 2024)

WEBSITES

www.kidshealth.org
Fun facts and games to learn more about your body.
Click on the section called 'for kids'.

https://www.dkfindout.com/uk/human-body/life-cycle
Information about the human life-cycle.

www.wonderopolis.org/wonder/why-do-some-kids-grow-faster-than-others
An article and activities about growing.

Note to parents and teachers:
Every effort has been made by the Publishers to ensure that
these websites are suitable for children, that they are of the
highest educational value, and that they contain no inappropriate
or offensive material. However, because of the nature of the
Internet, it is impossible to guarantee that the contents
of these sites will not be altered. We strongly advise that
Internet access is supervised by a responsible adult.